■ AT HOME WITH SCIENCE ■

A CHILLING STORY

HOW THINGS COOL DOWN

EVE & ALBERT STWERTKA
PICTURES BY MENA DOLOBOWSKY

JULIAN MESSNER

 Published by Julian Messner, a division of Silver Burdett Press, Inc., Simon & Schuster, Inc. Prentice Hall Bldg., Englewood Cliffs, NJ 07632.

JULIAN MESSNER and colophon are trademarks of Simon & Schuster, Inc.

Design by Malle N. Whitaker.
Manufactured in the United States of America.

Lib. ed.
10 9 8 7 6 5 4 3 2 1
Paper ed.
10 9 8 7 6 5 4 3 2 1

Library of Congress Cataloging-in-Publication Data
Stwertka, Eve.
A chilling story / Eve and Albert Stwertka.
p. cm. — (At home with science)
Includes index.
Summary: Discusses how refrigeration works.
1. Refrigeration and refrigerating machinery—Juvenile literature. 2. Air conditioning—Juvenile literature. [1. Refrigeration and refrigerating machinery.]
I. Stwertka, Albert. II. Title.
III. Series.
TP492.2..S78 1991 90-39293
621.56—dc20 CIP AC
ISBN 0-671-69457-X (lib. bdg.)
ISBN 0-671-69463-4 (pbk)

CONTENTS

CHILLING AND FREEZING

Many foods and drinks taste great when they are chilled. But there is a more important reason for chilling foods. Cold temperatures keep foods fresh much longer. People discovered this fact hundreds of years ago, but they didn't understand the reasons for it.

Today, we know that invisible germs and tiny **molds** live all around us. They settle in places where they can grow. The warmer the temperature, the more quickly germs multiply and spoil food. For example, in hot weather, milk turns sour. Bread turns moldy. Meat and fish spoil and can make us very sick.

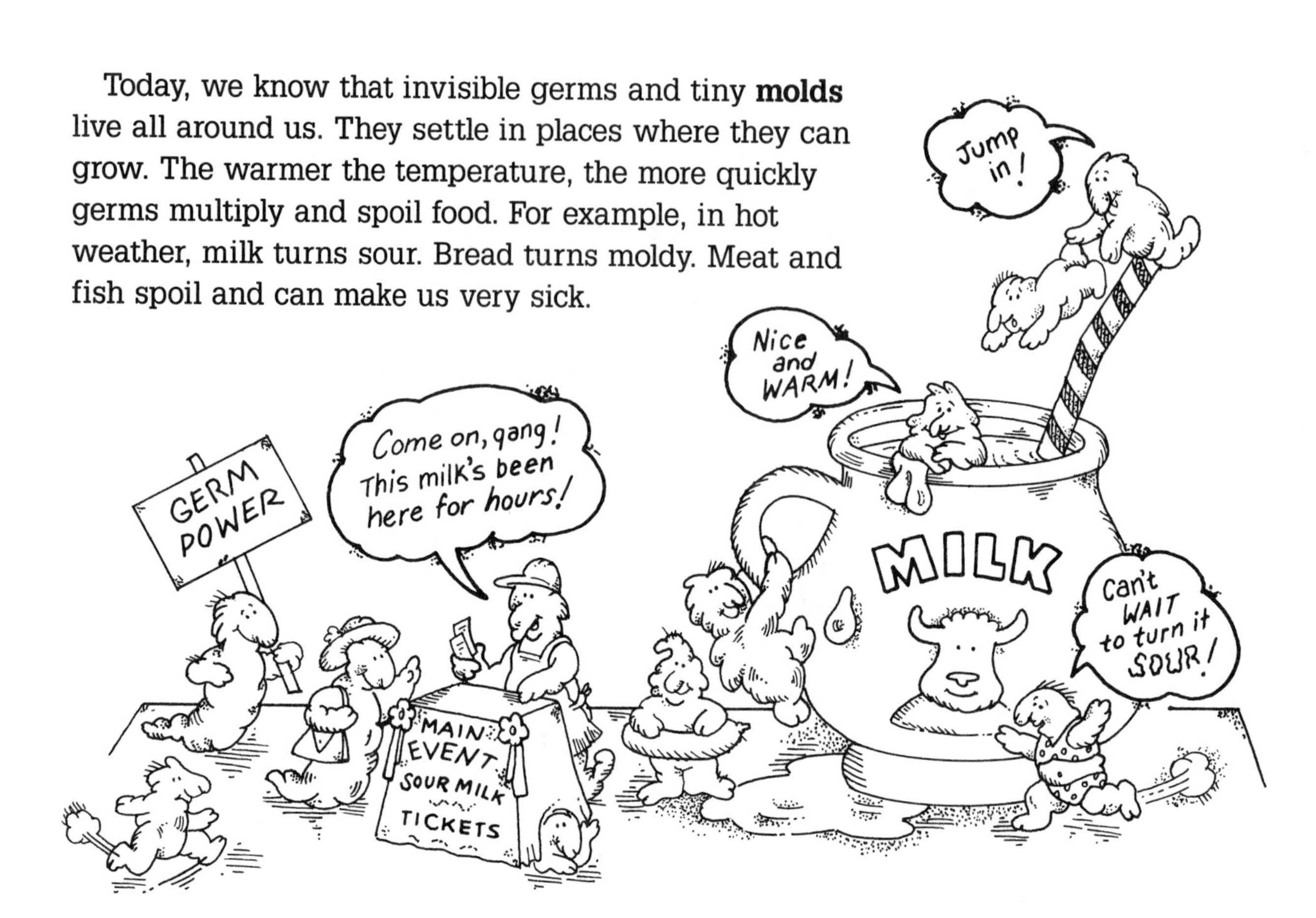

Try this...

Put two slices of cooked potato on two separate saucers. Cover them loosely with aluminum foil or wax paper. Place one slice in the refrigerator and the other in a closet or drawer. A week later, check both. You will see mold starting to grow on the warmer potato.

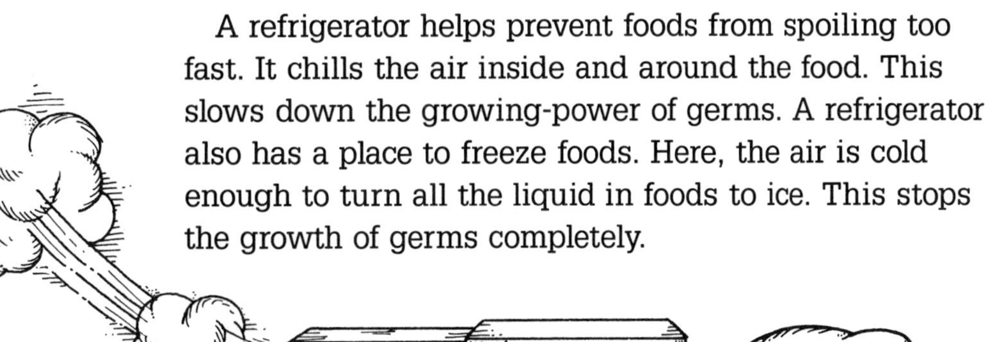

A refrigerator helps prevent foods from spoiling too fast. It chills the air inside and around the food. This slows down the growing-power of germs. A refrigerator also has a place to freeze foods. Here, the air is cold enough to turn all the liquid in foods to ice. This stops the growth of germs completely.

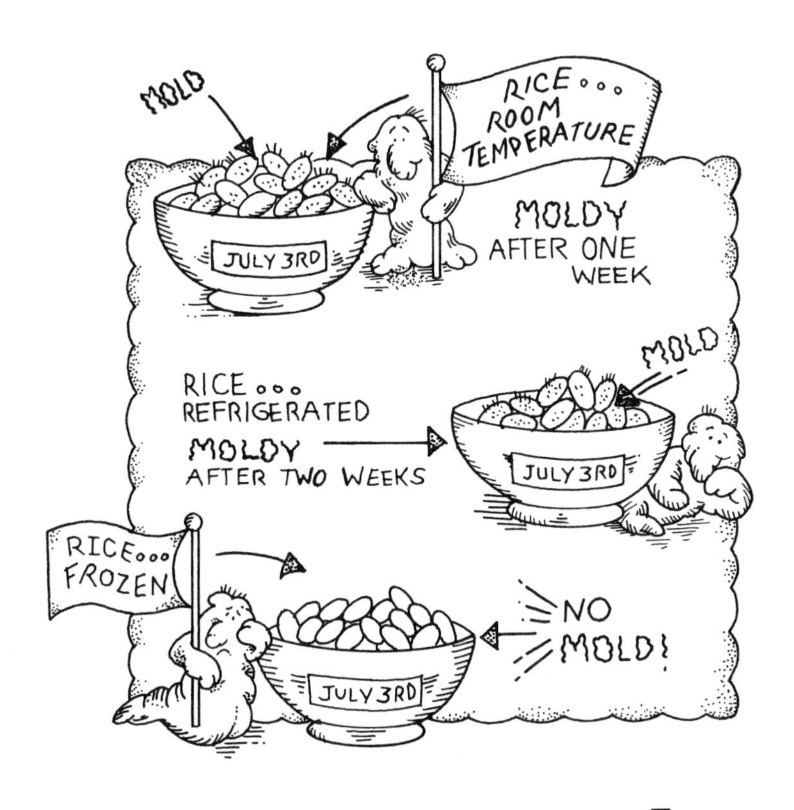

Try this...

Put a tablespoonful of plain, cooked rice in each of three covered containers. Label each container with the date. Then, store one container on a kitchen shelf. Put the second one inside the refrigerator. Place the third in the freezer.

Check the contents of each container after a week. You'll find that the rice kept on the kitchen shelf is the first to go moldy. Record the difference in the time it takes for the first two containers to show mold. As for the frozen rice, you can keep it for months and it will still be good to eat.

Ice is an amazing preservative. Explorers in the far north have found remains of whole animals frozen into cliffs of ice. One of these was a woolly mammoth—an early form of elephant that is now extinct. Although the frozen mammoth was thousands of years old, its meat was still fresh enough for the explorers' sled dogs to eat.

Well, you'll probably never need to store a woolly mammoth. But all the same, you can see that if you want to keep anything from spoiling for a long, long time, the best thing to do is to freeze it.

Even though refrigeration was not invented until the twentieth century, people quickly came to depend on it. Nowadays, we count on refrigerators and freezers to preserve foods and medical supplies.

Another cooling device we depend on is the air conditioner. This wonderful invention keeps people comfortable in hot weather. Some kinds of electronic machines need to be kept cool. Large computers, for example, are often kept in air-conditioned rooms.

Think of all the ways we use refrigeration. See if you can add to this list.

- To freeze ice cream
- To cool trucks that bring milk from the farm
- To cool railroad cars bringing Thanksgiving turkeys to the supermarket
- To make an ice skating rink
- To keep you cool at the movies
- What else?

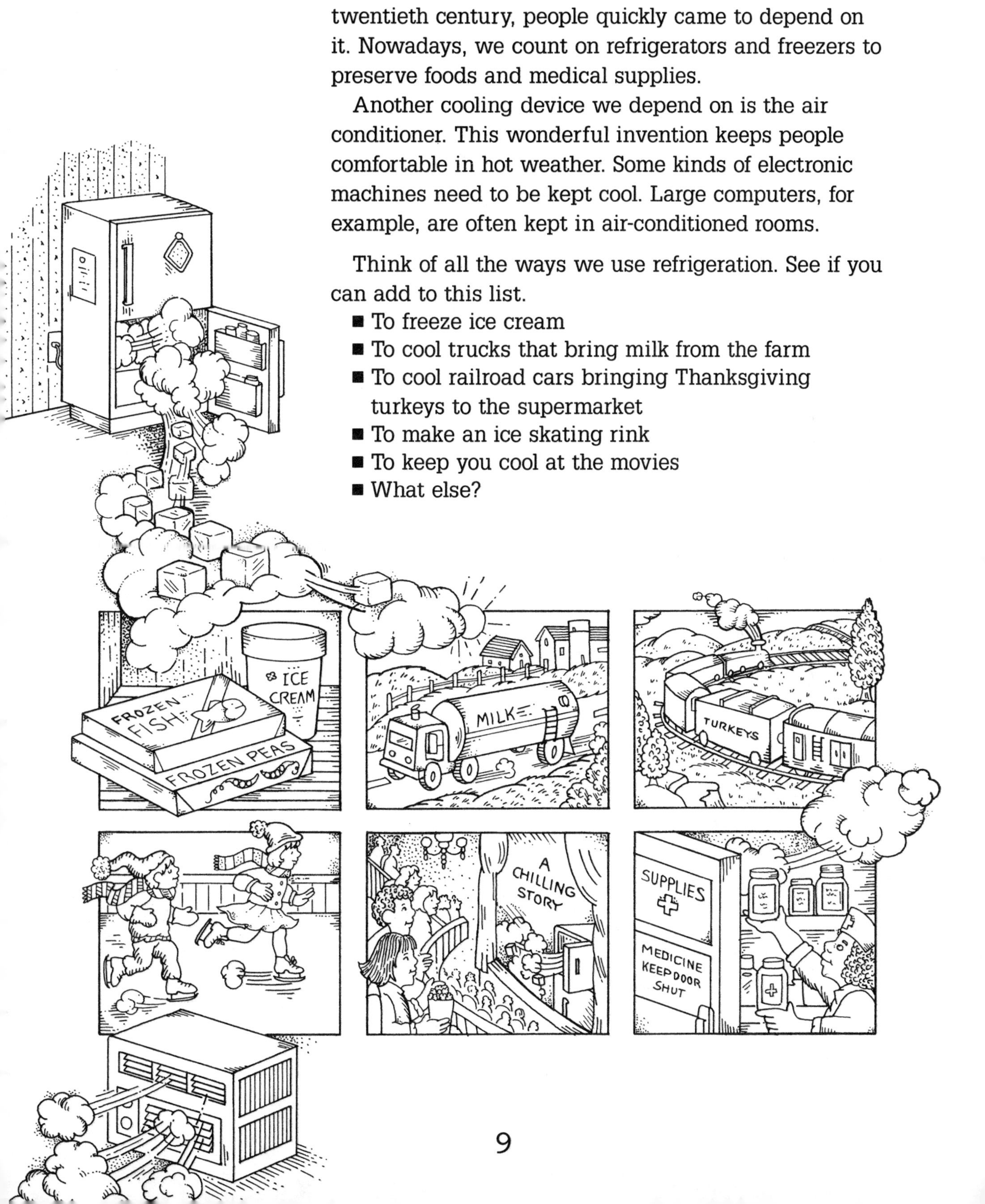

THREE WAYS TO COOL THINGS DOWN

Cooling By Heat Transfer

Remember how it feels when you lick an ice cream cone? Your tongue turns cold, and the ice cream turns warm. Heat from your tongue flows to the cold ice cream and warms it until it melts.

Heat naturally flows from warm things to cold things. In fact, to cool something simply means to take heat away from it. The movement of heat from your tongue to the ice cream is called **heat transfer.** Heat transfer is one of the ways we make things cold.

It is much harder to make things cold than to make them hot. If you want to boil water, you can start a fire and heat a pot of water over it. But how would you freeze water on a warm day?

To make even a single ice cube without the help of cold weather, you need that wonderful modern invention—the refrigerator.

Try this...

Drop an ice cube into a small glass of warm water. Notice how the heat in the water transfers to the ice cube and melts it. At the same time, the water cools.

In ancient times, people knew about cooling by heat transfer. They used this method to preserve food and drinks. In the winter, they shoveled snow into underground caves and packed it against the floor, walls, and ceiling. Protected from the sun by rocks and earth, the snow melted very slowly. Foods tucked away in this chilly space stayed fresh well into the summer.

More recently, in your great-grandparents' days, blocks of ice were cut from frozen lakes in winter and stored underground. Trucks brought the ice to people's houses, where an icebox stood in every kitchen.

In hot weather, the iceman came every day to push a hefty block of ice into the top part of the box. The melting ice drained through a tube into a basin on the floor.

An icebox works by heat transfer. The air inside the box transfers its heat to the ice. The stored food transfers its heat to the chilled air around it. The warmer the air, the more heat is transferred to the ice and the faster the ice melts.

Cooling by Evaporation

The moment you step out of a bath or shower, you tend to feel cold. If you don't have a towel handy, you'll probably shiver for a while. Soon, though, you will be perfectly dry. What happened to the water on your skin? It turned into a gas called water **vapor.** Through the process of **evaporation,** the water disappeared into the air.

Water vapor is hard to imagine because you can't see it or smell it. Notice what happens, though, when you open a bottle of perfume. Your nose can tell right away that something is there. Perfume is scented water or alcohol. When the liquid evaporates, the scent rises with the vapor and stays in the air.

Water evaporates by taking in heat from its surroundings. You feel cold when you are wet because the evaporating water is drawing some of the heat away from your skin. Evaporation, then, is a second way to cool things down.

Before air conditioning was invented, people used fans to cool themselves. A fan moves the surrounding air and helps the slight moisture on your skin to evaporate more quickly.

Try this...

Hold up your two index fingers. They probably feel about the same temperature. Now wet one finger and hold them both up again. Because the water is evaporating, the wet finger feels much colder than the dry finger.

Now wet your finger again and wave it around. The air rushing by speeds up evaporation and makes the finger feel even colder than before.

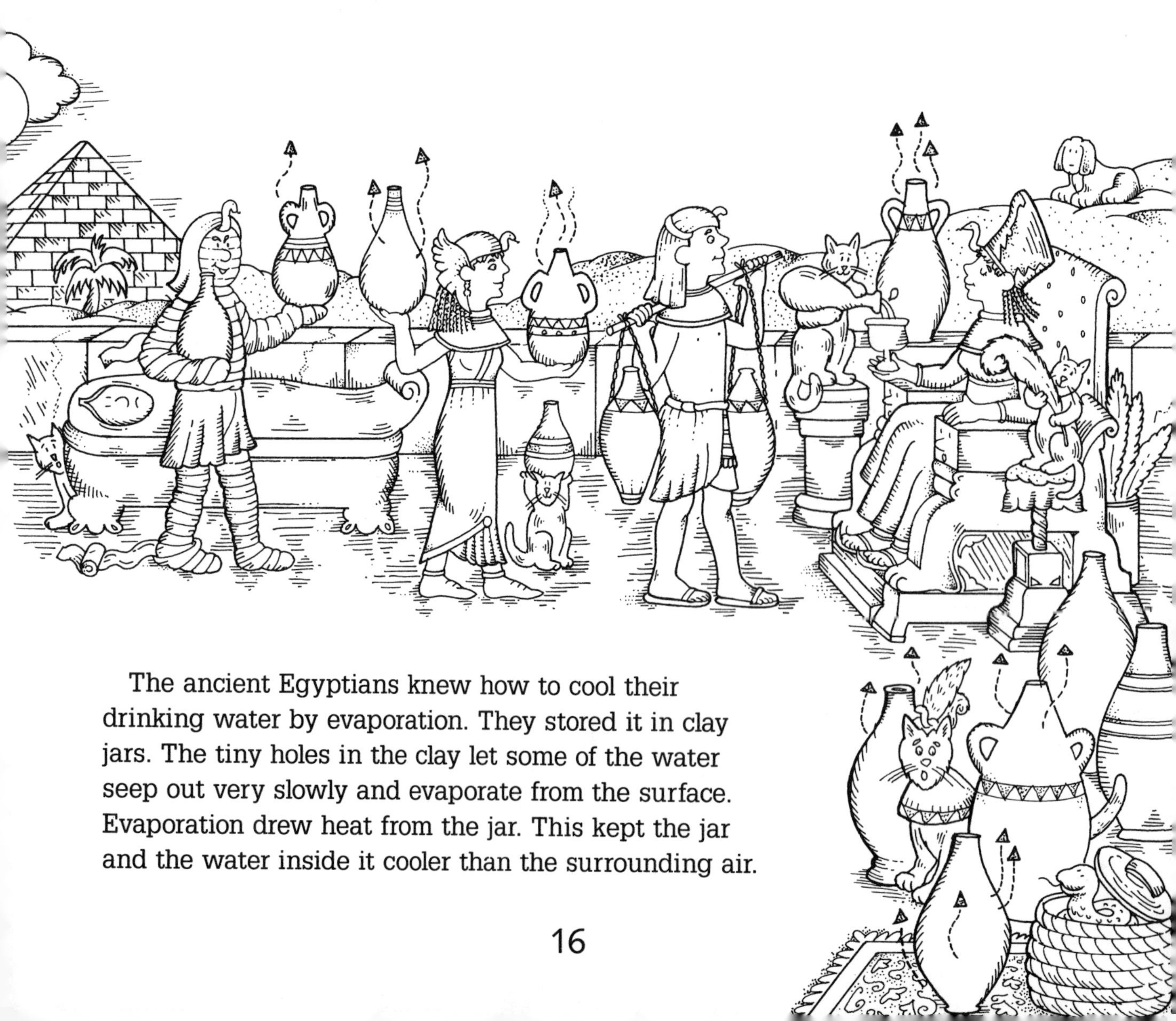

The ancient Egyptians knew how to cool their drinking water by evaporation. They stored it in clay jars. The tiny holes in the clay let some of the water seep out very slowly and evaporate from the surface. Evaporation drew heat from the jar. This kept the jar and the water inside it cooler than the surrounding air.

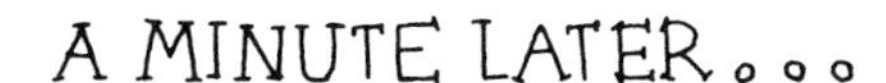

When water vapor touches something colder than itself, it loses heat and can't remain a gas any longer. It has to return to its liquid state and become water again. This is the opposite of evaporation. We call it **condensation.**

Condensation is easy to watch. Fill a dry glass with ice cubes and set it down. Soon the outside of the glass becomes coated by beads of water. This liquid comes from water vapor in the air. The vapor is forced to condense as soon as it touches the cold glass.

Cooling by Expansion

Have you ever used a pump to blow up a bicycle tire? If so, you may have noticed that the pump soon starts to feel warm. The reason is that you're putting pressure on the air in the pump every time you push down the plunger.

When air or other gases are squeezed together they turn hot. This process is called **compression.** On the other hand, when gases are released and allowed to expand, they turn cold. This process, called **expansion,** is still another way to make things cool.

Try this...

Blow up a balloon. Now open it and feel the escaping air. The skin of the balloon has been compressing the air inside. As the air rushes out it expands and cools.

Or try this...

Blow on your hand. The air coming from your lungs feels warm, like the temperature inside your body.

Next, make your mouth smaller by puckering your lips. This time, when you blow on your hand it feels much cooler, doesn't it? That's because you compressed the air inside your mouth. As soon as it escaped your lips, though, the air expanded and turned colder.

Now you know three different ways to cool things down:

1. By HEAT TRANSFER
2. By EVAPORATION
3. By EXPANSION

Put them all to work together, and you have the secret of refrigeration.

BEHIND REFRIGERATOR WALLS

Open the door of your refrigerator. What do you see? Food, of course, and bottles and jars. But aside from these, there seem to be only shelves and storage bins.

To make the refrigerator easy to use and easy to clean, most of its working parts are hidden. If you look closely, though, you will see small vents in the inside walls. You might also be able to see the fan that moves air through the vents so that every part of the refrigerator is cooled. Notice how thick the walls and the doors of the refrigerator are. They are filled with a material to keep the cold air in and the warm air out. This barrier slows down the transfer of heat. It is called **insulation.**

Try this...

Take two ice cubes of the same size from the freezer. Wrap one of them quickly and tightly in a washcloth and fasten it with a rubber band.

Put the second cube on a dish. Set them down side by side. When the open ice cube has just about disappeared, unwrap the insulated one. Most of it will still be there. Record the difference in the time it takes for the two cubes to melt.

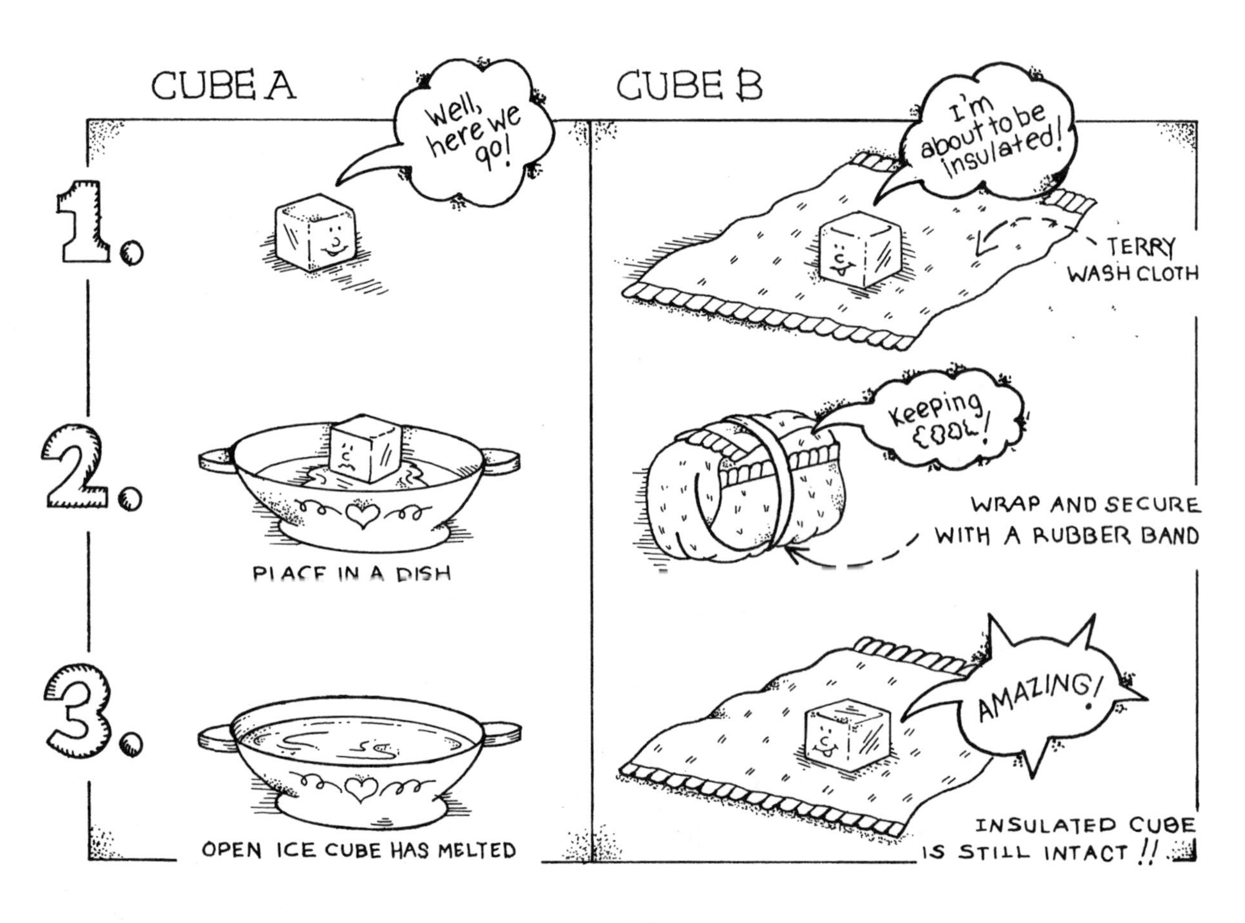

The heart of the refrigerator is the **compressor.** This is what makes the noise you hear when the refrigerator is working. The compressor acts like a pump. It pumps a fluid called a **refrigerant** through the various parts of the refrigerator.

The refrigerant absorbs heat from inside the refrigerator and transfers it to the air outside. It does this over and over again as it passes through the system.

The **refrigeration cycle** begins when the liquid refrigerant is forced at high pressure through a tiny channel called an **expansion valve.** The valve opens into a coil of metal tubing with many turns. Inside this coil, the refrigerant is free from pressure. As soon as the liquid is released into the coil, it evaporates into a gas and expands. In the process, it loses heat and turns cold.

The coil where the refrigerant evaporates is called the **evaporator.** Swirling around the cold evaporator coil, the air in the refrigerator gives up heat and turns colder. Drops of water from the moisture in the air condense on the metal. A tube at the back of the freezer lets this water drip down to a pan on the floor.

Look behind the grate at the bottom of your refrigerator. Can you find the water pan? Can you see a small fan that moves the air to help this water evaporate and not spill over?

When your kitchen is warm and humid, you can see the water vapor in the air condense into tiny droplets the moment it enters the cold refrigerator. Open the freezer door and watch the entering air turn into a cloud of white mist.

Sometimes the water that forms on the evaporator coil freezes. It turns into tiny ice crystals called frost. If frost builds up on the evaporator coil, it can block the cool air from circulating properly. Most refrigerators have a small heating coil near the evaporator. This coil turns on at certain times to melt, or "defrost" the unwanted ice.

Listen to your refrigerator. Can you tell when the defrosting is taking place? You'll hear gurgles and noises that sound like ice cracking.

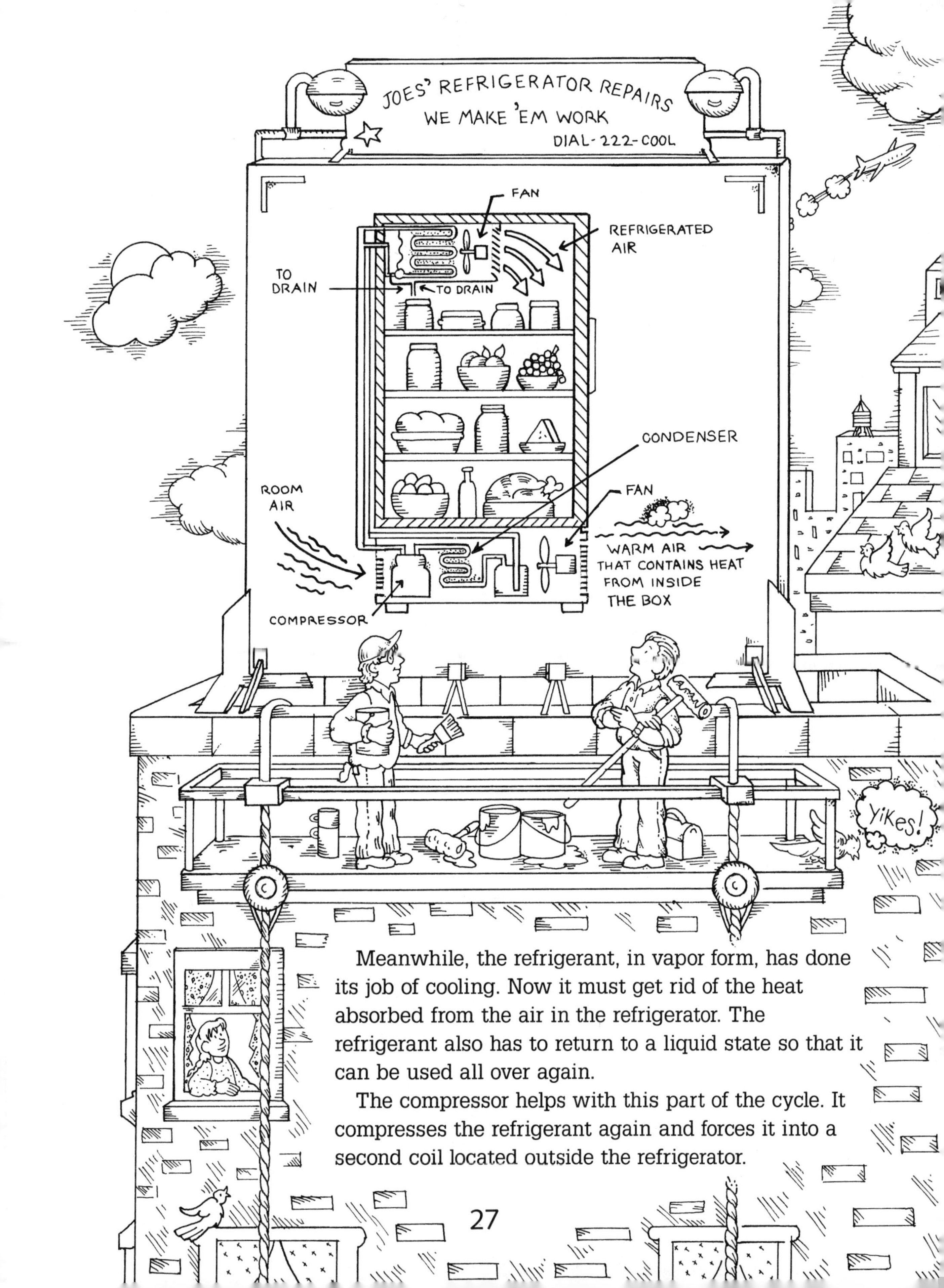

Meanwhile, the refrigerant, in vapor form, has done its job of cooling. Now it must get rid of the heat absorbed from the air in the refrigerator. The refrigerant also has to return to a liquid state so that it can be used all over again.

The compressor helps with this part of the cycle. It compresses the refrigerant again and forces it into a second coil located outside the refrigerator.

Made even hotter by compression, the refrigerant gas circulates through this outside coil, called the **condenser.** Because the outer air is cool by comparison even on the hottest day, the refrigerant loses heat. It condenses and turns liquid once again.

At last, from the condenser, the liquid flows back to the evaporator and starts the entire cycle again.

The condenser is usually located at the rear of the refrigerator. Put your hand back there and see if the air feels warm.

The food compartment is usually kept at a temperature a little above freezing. If it got too cold, all the food would freeze. To stop the refrigerator from becoming too cold, a device called a **thermostat** is used. When the refrigerator gets too cold, the thermostat acts like a switch and turns the compressor off. This stops the refrigeration cycle. When the refrigerator warms up, the thermostat turns the compressor on so that cooling can start again.

Listen to your refrigerator. When it is quiet, the thermostat has turned off the compressor.

HOW AIR CONDITIONING WORKS

Air conditioning has made summer more comfortable and more fun. We can sleep better on hot nights, go to the movies on a scorching day in August, or take long rides in cars, buses, or trains without feeling as though we are in an oven.

An air conditioner is very similar to a refrigerator. But an air conditioner cools a much larger space.

Like a refrigerator, an air conditioner has a compressor that circulates a refrigerant through the cooling system. A fan takes in warm air from the room, cools it by passing it over the evaporator coils, and blows it back into the room again. The refrigerant then carries the heat from the room to the condenser, which is located at the back of the air conditioner and outside the house. Meanwhile, the compressor has made the refrigerant gas even hotter, so that heat is removed by the surrounding air, even on warm summer days. Then the cycle repeats itself.

When the warm air from the room passes over the cooling coils, much of the moisture it contains condenses and turns to water. This water drains to the back of the air conditioner. There it can evaporate into the outside air or drip down to the ground.

In hot, humid weather, look at the ground outdoors under an air conditioner. Notice how much water has been removed from the room.

HOME *WITHOUT* AIR CONDITIONING...

HOME *WITH* AIR CONDITIONING...

Removing water vapor, or dehumidifying the air, is one of the important jobs an air conditioner can do. We usually feel cooler and more comfortable in dry air because it lets our perspiration evaporate more easily than humid air.

Have you ever noticed the fresh smell of an air conditioned room? Air conditioning actually cleans the air. As it removes moisture from the air, it also removes dust particles, pollens, and pollutants that cling to the condensing water droplets. That's why people who suffer from hay fever and other allergies are less apt to cough and sneeze in an air conditioned room or car.

Air conditioning is important for the complicated electronics used in computer centers, laboratories, and hospitals. Metal and other materials expand and contract as they get warm and cold. If this kind of equipment is not kept at a constant cool temperature, its delicate parts stop working properly.

DANGER: FREON AT WORK

You may be wondering about the magical substance that travels round and round the refrigeration cycle, growing cold and hot and cold again. It is a chemical compound called **Freon.** Freon makes a perfect refrigerant. It turns easily from a gas into a liquid and back again at just the right temperatures.

Freon also seems perfectly safe. If it escaped from a leaky refrigerator coil, it would not damage the food, harm the people breathing it, or catch fire.

Recently, though, scientists have discovered that Freon is harmful to the environment. When Freon escapes from leaking refrigerators and home or car air conditioners, it slowly rises high above the earth. Here, it reduces the thickness of the **ozone layer,** the upper part of the **atmosphere.** The ozone layer protects people, plants, and animals from some of the dangerous rays given off by the sun.

You can see how important it is to keep Freon out of the air. Today, engineers are developing new refrigerants that will not damage our environment. In years to come, people will be able to enjoy the benefits of cooling power, knowing that the layer of air around the earth is being kept safe and sound.

CHILLING WORDS

Atmosphere (AT-mas-feer)—the layer of gases that surrounds the earth.

Compression (kom-PRES-shun)—the act of forcing or pressing something into a smaller space. When air and other gases are compressed, they become hot.

Compressor (kom-PRES-er)—a kind of pump in the cooling system that forces the refrigerant through the refrigeration cycle.

Condensation (kon-den-SAY-shun)—the change that occurs when a gas or vapor turns into a liquid.

Condenser (kun-DEN-sir)—a device in the cooling system that transfers heat out of the system by changing the refrigerant vapor to a liquid state.

Evaporation (ee-vap-uh-RAY-shun)—the change that occurs when a liquid turns into a gas or vapor.

Evaporator (ee-VAP-uh-ray-ter)—a device in the cooling system in which the refrigerant evaporates while absorbing heat.

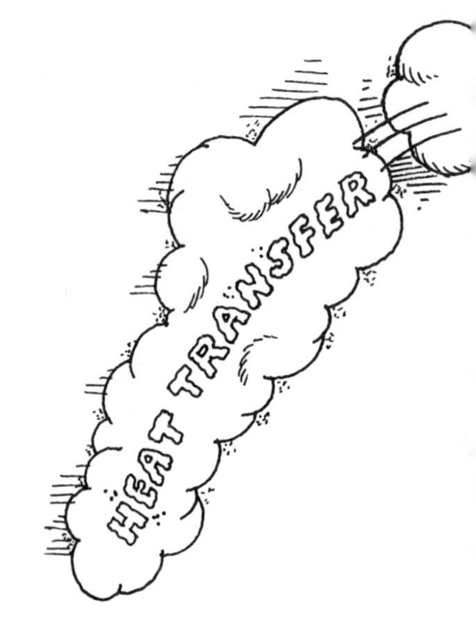

Expansion (ik-SPAN-shun)—an increase in size or volume. When air or other gases are allowed to expand, they lose heat.

Expansion valve—a device in the cooling system that helps turn the liquid refrigerant into a vapor.

Freon (FRE-on)—a chemical substance used as a refrigerant. Though harmless in many ways, it has recently been found to damage the ozone layer of the atmosphere.

Heat transfer—the flow of heat from warm things to cold things.

Icebox—the ancestor of the refrigerator. It cooled food with blocks of ice brought from outside.

Insulation—a material that keeps heat from flowing into the cold parts of the refrigerator.

Mold—a tiny life form that grows on animal or vegetable matter.

Ozone layer—a part of the atmosphere around the earth that protects us from the dangerous rays of the sun.

Refrigerant (re-FRIJ-uh-rent)—the chemical substance, such as Freon, that circulates around the refrigeration system.

Refrigeration cycle (re-FRIJ-uh-RAY-shun SI-kul)—the repeated changes of the refrigerant from a liquid to a gas and back again, as it removes heat from the refrigerator box and transfers it to the outside air.

Thermostat (THUR-muh-stat)—a device used to maintain a constant temperature.

Vapor (VA-pur)—the gaseous state of many liquids.

INDEX